Counting Cosmic Robots:
A GOOFY GALACTIC ADVENTURE

A MISTER JON BOOK

ISBN: 9798858436591
IMPRINT: INDEPENDENTLY PUBLISHED

FOR ZACK

Dear Friends,

In the realm of early learning, where curiosity flourishes, a solid foundation is key. "Counting Cosmic Robots" is a vibrant bridge to this world, offering young minds the excitement of exploration wrapped in the joy of numbers. Through its playful characters and interactive tales, this book isn't just about counting – it's about fostering a lifelong love for learning.

Each page brims with colorful robots and celestial wonders, creating a delightful space for children to not only count but also engage, giggle, and discover. The journey through these pages isn't just educational; it's an adventure that equips children with essential cognitive skills while making learning an enjoyable part of life.

As you open the pages of "Counting Cosmic Robots" remember that you're igniting a spark that will illuminate their educational path. This book isn't just a story; it's a tool for early learning, a ticket to laughter, and an opportunity to plant the seeds of curiosity that will flourish as they grow.

Cheers,
Mister Jon
Panajchel, Guatemala

ROUND ONE
WARM UP

ONE

TWO

THREE

FOUR

FIVE

ROUND COMPLETE
GOOD JOB

ROUND TWO
QUICK ROUND

Robot number one, a curious fellow,
With eyes that gleam, orange, warm, and mellow.

Robot number two, with giggles so bright,
Twirls with the stars in the cosmic night.

Robot number three, a comet in flight,
Zipping and zooming, pure delight.

Robot number four, like a bouncy ball,
Hops on meteors, having a ball.

Robot number five, a sparkle in space,
Spins through nebulas, a colorful chase.

Robot number six, galaxies it spins,
Whirling through stars with cosmic grins.

Robot number seven, on moons it prances,
Bouncing and laughing, cosmic dances.

Robot number eight, a star so bold,
Shimmering and shining, a sight to behold.

Robot number nine, like a comet's blaze,
Zips and zooms through cosmic ways.

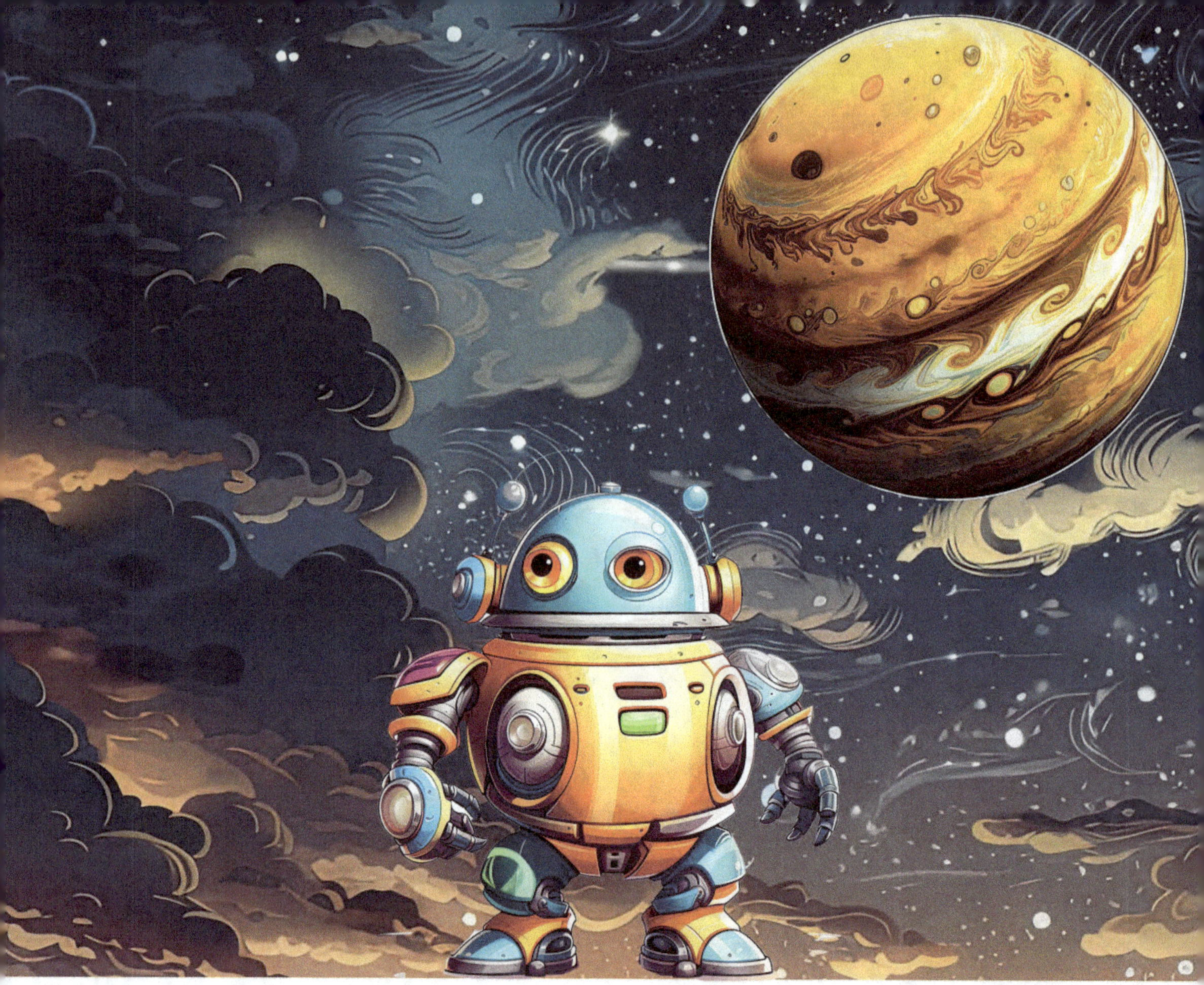

Robot number ten, a planet so grand,
Dances in orbits across the land.

10

ROUND COMPLETE
GOOD JOB

MAIN
ROUND

Greetings from
Florida

In a galaxy so whimsical and bright,
Twenty Gigglebots embarked one night.
With googly eyes and laughter to treasure,
They counted the cosmos, spreading joy without measure.

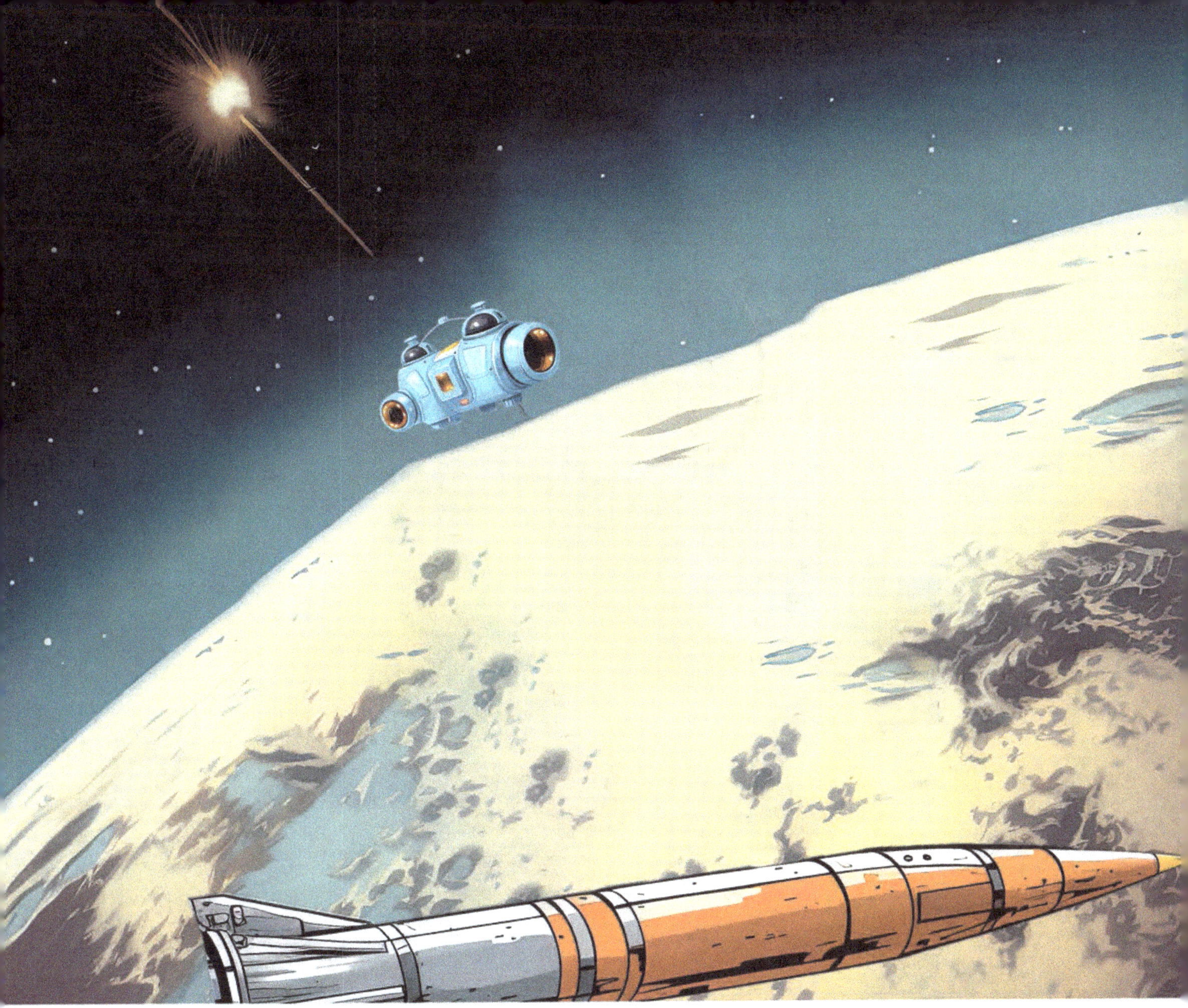

Boopatron counted one twinkling star,
With antennas wiggling, he traveled far.

Sparklebot saw two fluffy clouds,
Spreading giggles, and laughs out loud.

Bloopzilla scanned three wobbling moons,
In a cosmic dance, they sang silly tunes.

Zoombo whizzed around four silly suns,
His laughter echoed, a cosmic fun.

Luna-Doodle saw five meteors soar,
Bouncing with glee, SHE wanted more.

Galaxi-Giggle beamed with glee,
Six polka-dot galaxies she did see.

Stardust Sparklepants spread delight,
Seven cosmic rainbows, colors bright.

Aurora Borealis danced with glee,
Eight celestial beings, a sight to see.

8

Nebula Noodle, full of spunk,
Nine space donuts, a cosmic dunk.

Comet Cruncher crunched ten intergalactic cookies,
Leaving trails of crumbs, oh my, how kooky!

10

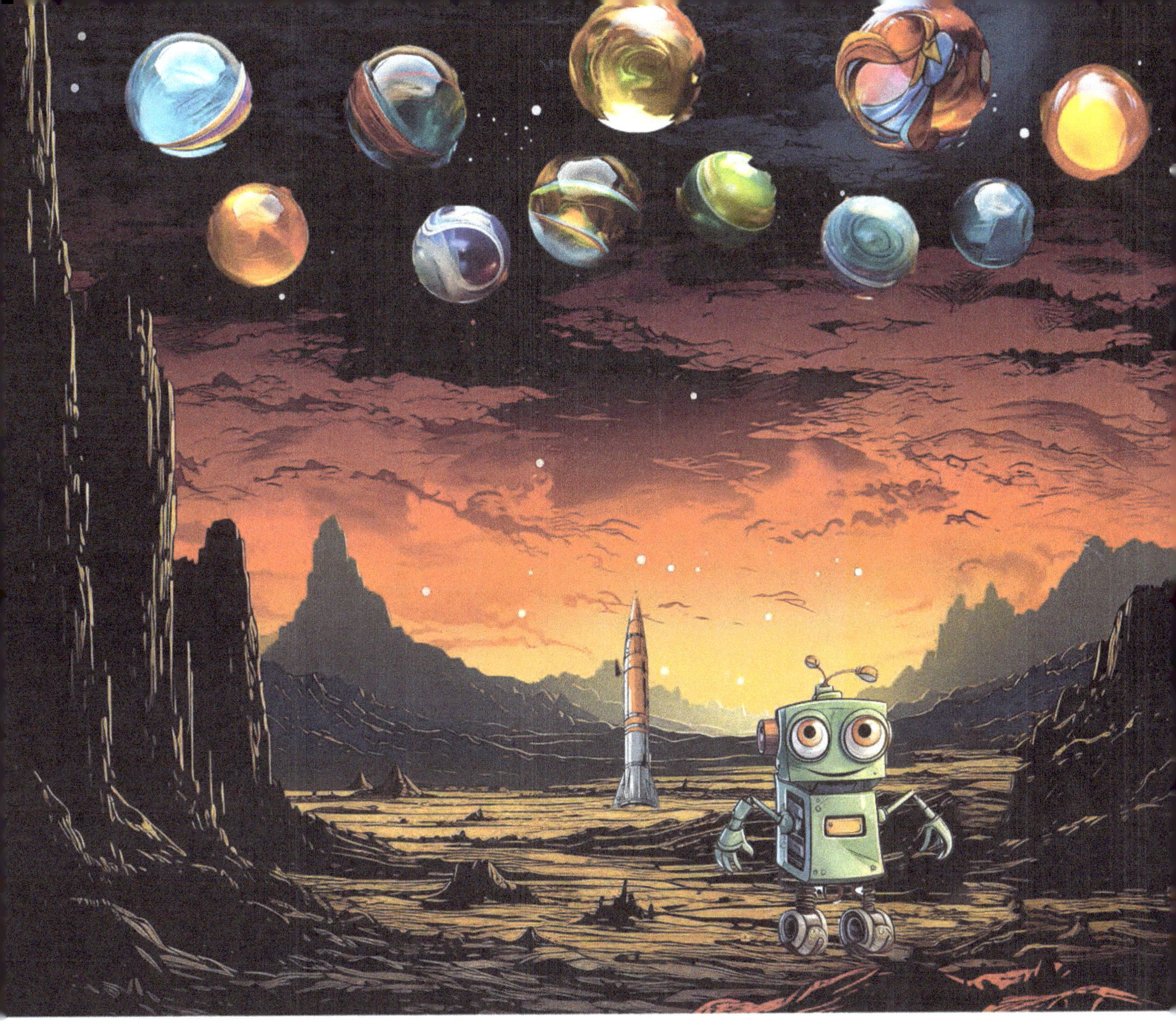

Fluffybot, oh what fun,
Eleven candy planets, yum yum!

11

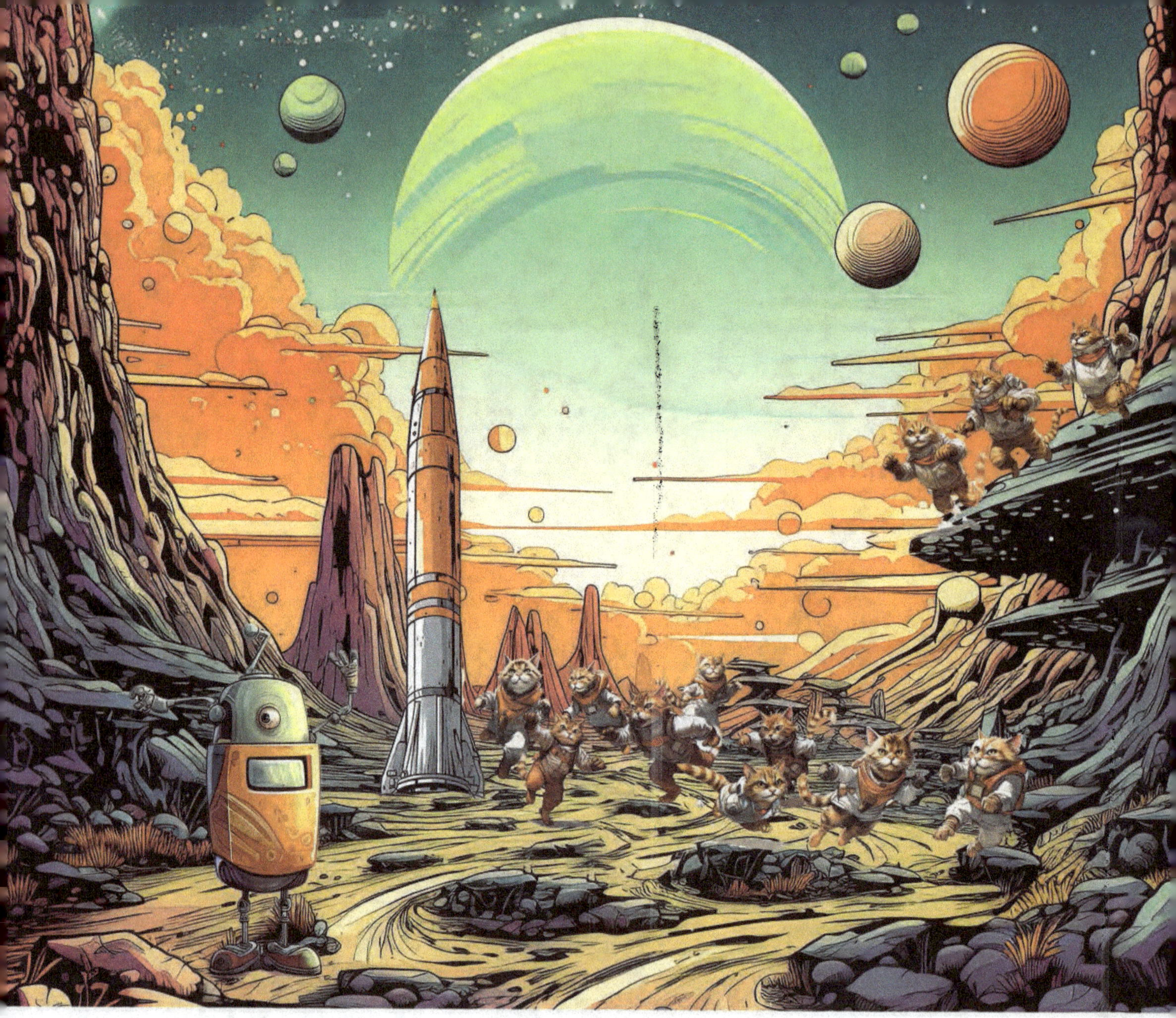

Quirkatron counted twelve cosmic cats,
as they raced, and chased through the flats.

12

Wobblewiggles floated past thirteen
space jellyfish so fair,
Their wibbly-wobbly dance, a cosmic sight rare.

Astro-Giggles gleaming, with a cosmic smile so vast,
Fourteen aliens waving, having a space blast!

14

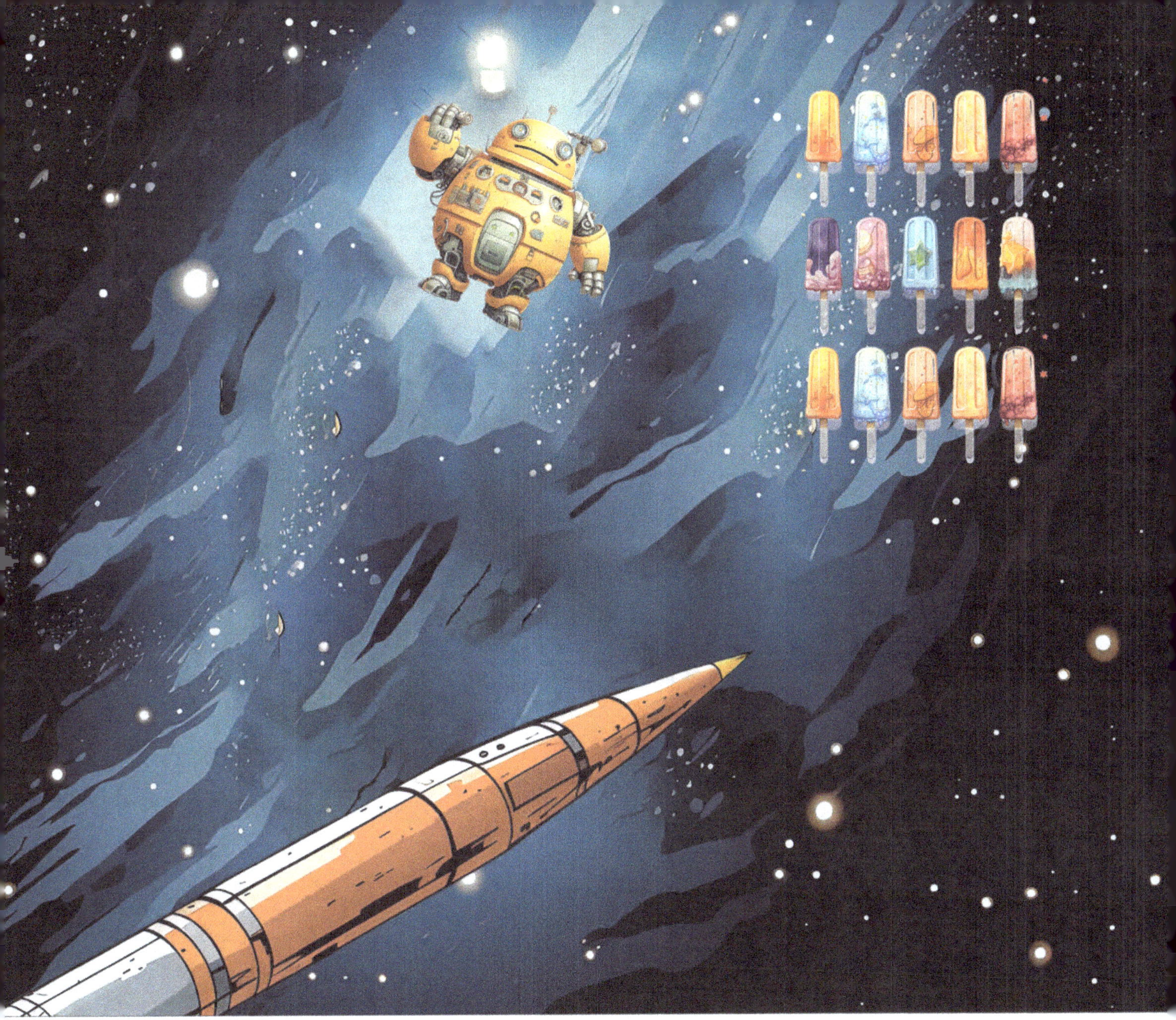

Starburster beamed fifteen cosmic popsicles,
With flavors galore, a celestial spectacle.

15

Meteor Muncher gobbled sixteen cosmic cupcakes,
Leaving no crumbs, not even a flake.

16

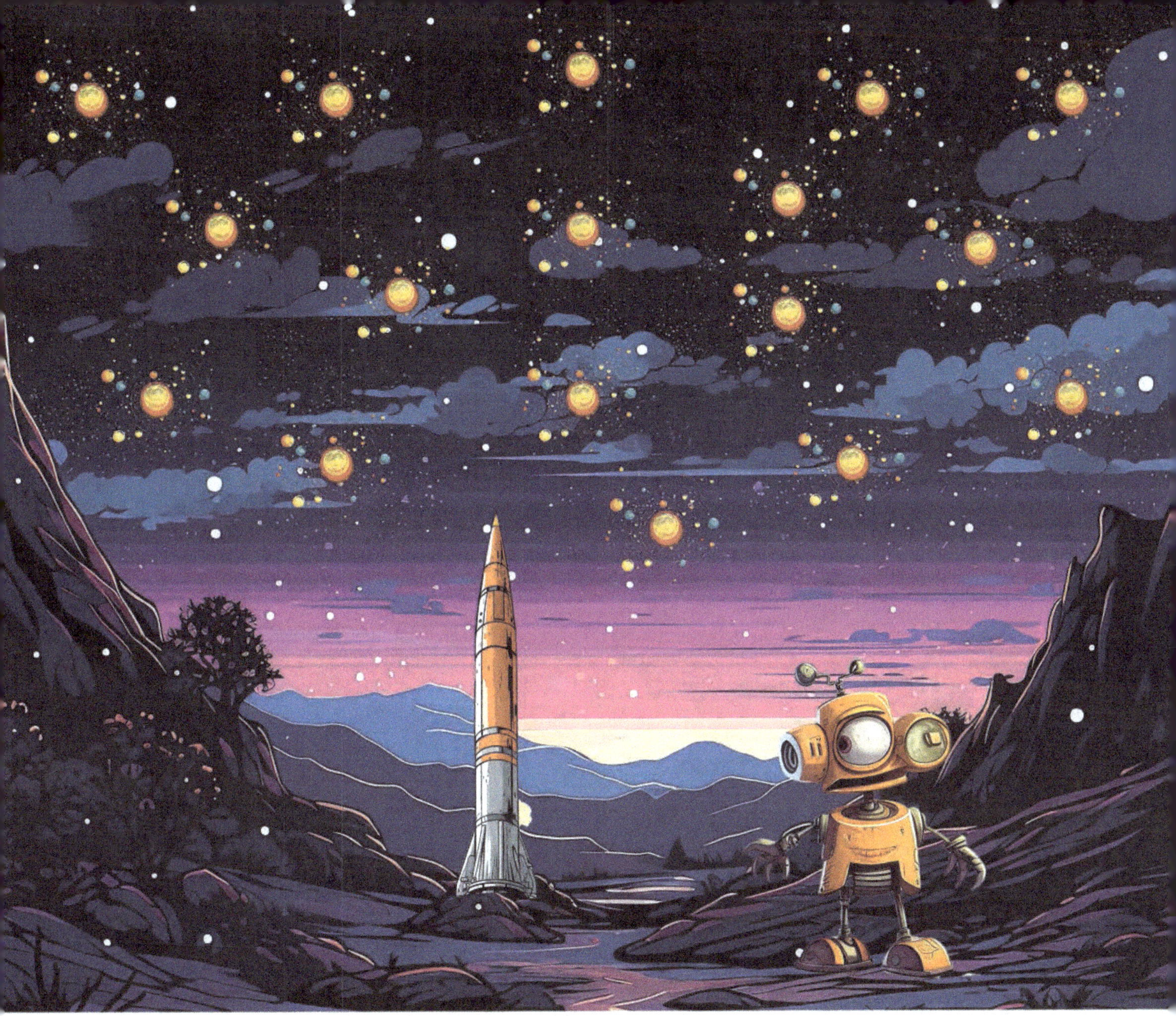

Giggleplex explored seventeen giggle galaxies,
In a universe of laughter, he was never lazy.

17

Warp-Wooble ventured, full of glee,
Through eighteen space tacos, a cosmic spree.

Zippy-Zonk raced 'neath nineteen stars so grand,
Their twinkling lights, a cosmic wonderland.

19

Gigglopolis marveled at twenty silly supernovas,
In a galaxy of laughter, they lived happily ever after.

Hand in hand, these bots did traverse,
Through the cosmic expanse, a joyful universe.
Counting celestial wonders, with laughter and glee,
Their cosmic journey, a wondrous spree.

ROUND COMPLETE
GOOD JOB

BONUS ROUND

Can you spot the 3 robots in the cosmic parking lot?

Can you find the 3 robots in the factory?

KNOW YOUR ROBOT

Boopatron
Hobbies: Doing the robot dance, telling jokes, and making silly faces.
Favorite Food: Bubblegum-flavored bolts and nuts.
Favorite Color: Goofy Green.

Bloopzilla
Hobbies: Creating fizzy cosmic concoctions, riding rainbows, and tap-dancing on asteroids.
Favorite Food: Giggling grape jelly sandwiches.
Favorite Color: Sparkly Pink.

Zoombo
Hobbies: Buzzing around like a bee, juggling rubber chickens, and playing cosmic pranks.
Favorite Food: Banana pudding with marshmallow sprinkles.
Favorite Color: Wacky Yellow.

Sparklebot
Hobbies: Whizzing through time and space, building silly contraptions, and playing with shiny marbles.
Favorite Food: Whirly-popcorn with extra giggles.
Favorite Color: Whimsical Orange.

Galaxi-Giggle
Hobbies: Zipping through the stars, playing hide-and-seek with comets, and breakdancing on meteoroids.
Favorite Food: Zany zigzagging zucchini.
Favorite Color: Bopping Blue.

Stardust Sparklepants
Hobbies: Wiggling like a jellybean, making cosmic cupcakes, and telling knock-knock jokes to planets.
Favorite Food: Giggly Gummy Worms.
Favorite Color: Wiggly Purple.

Luna-Doodle
Hobbies: Snickering with laughter, playing cosmic charades, and doing the robot hokey-pokey.
Favorite Food: Silly S'mores with marshmallow mustaches.
Favorite Color: Chuckle Brown.

Aurora Borealis
Hobbies: Spreading cosmic joy, telling punny riddles, and doing the hula dance with asteroids.
Favorite Food: Jolly Jello Jigglers.
Favorite Color: Laughing Lavender.

Comet Cruncher
Hobbies: Bouncing through space,
bouncing cosmic boulders, and
bouncing jokes off satellites.
Favorite Food: Bubbly bouncing
blueberries.
Favorite Color: Bouncy Turquoise.

Fluffybot
Hobbies: Getting dizzy with laughter,
juggling planets, and tap-dancing
with meteor showers.
Favorite Food: Fizzy fizzy fruit punch.
Favorite Color: Spinning Cyan.

Nebula Noodle
Hobbies: Snazzing up the galaxy,
dancing the tango with stars,
and hosting cosmic fashion shows.
Favorite Food: Snickerdoodle
sparkles.
Favorite Color: Tickle Pink.

Quirkatron
Hobbies: Creating whimsical wonders,
doing the hula-hoop with moons,
and surfing on cosmic waves.
Favorite Food: Jiggly jellybean jubilee.
Favorite Color: Whirling Rainbow.

Meteor Muncher
Hobbies: Snorting cosmic stardust, playing hide-and-seek with galaxies, and doing the robot twist.
Favorite Food: Snickering Snickerdoodles.
Favorite Color: Wiggly Teal.

Starburster
Hobbies: Chuckling with cosmic delight, bouncing on moonbeams, and juggling space bananas.
Favorite Food: Chuckleberry cheesecake.
Favorite Color: Fizzing Magenta.

Astro-Giggles
Hobbies: Giggling like a cosmic hyena, hosting wacky talent shows, and spinning cosmic whirligigs.
Favorite Food: Giggling grapes with silly sprinkles.
Favorite Color: Whirly Yellow.

Wobblewiggles
Hobbies: Wearing mismatched socks, doing the robot mambo, and painting funny faces on asteroids.
Favorite Food: Wacky watermelon waffles.
Favorite Color: Bopple Red.

Giggleplex
Hobbies: Tickle-tag with comets,
hopping like a kangaroo on
asteroids,.
Favorite Food: Tickleberry taffy
treats.
Favorite Color: Giggling Coral.

Gigglopolis
Hobbies: Bopping with cosmic beats,
inventing silly gadgets, and playing
cosmic tag with moons.
Favorite Food: Bopcorn with
sprinkles of joy.
Favorite Color: Chuckleberry Blue.

Zippy-Zonk
Hobbies: Whimsical wand-waving,
doing the robot hokey-pokey,
and riding giggling shooting stars.
Favorite Food: Whimsy-whirl
ice cream.
Favorite Color: Bouncing Aqua.

Warp-Wooble
Hobbies: Giddy-up like a cosmic
cowboy, playing hopscotch on
planets, and telling cosmic puns.
Favorite Food: Giddy Gummy
Galaxies.
Favorite Color: Snickerdoodle Yellow.

Sparkle Gears Spark
Hobbies: Polishing Moon Rocks,
Creating Constellation Art,
Hosting Space Tea Parties
Favorite Food: Cosmic Cupcakes
with Stardust Sprinkles
Favorite Color: Twinkling Silver

Zoom Zoomington
Hobbies: Racing Shooting Stars,
Exploring Meteor Showers,
Chasing Comet Tails
Favorite Food: Rocket Fuel Smoothies
Favorite Color: Meteorite Red

Giggles McSprocket
Hobbies: Telling Cosmic Jokes,
Making Nebula Noodles,
Organizing Dance-Offs
Favorite Food: Galactic Gummy
Bears
Favorite Color: Rainbow Spectrum

Blinky Buzz Whirly
Hobbies: Doing the Robot Boogie,
Collecting Shooting Stars, Telling
Jokes Only Robots Understand
Favorite Food: Quantum Quinoa
Quesadillas
Favorite Color: Electric Blue

Sparky Zaptron
Hobbies: Creating constellations, inventing wacky gadgets, hosting interstellar picnics
Favorite Food: Solar-flare popcorn
Favorite Color: Electric blue

Giggles Chucklebot
Hobbies: Organizing laugh-a-thons, painting nebula masterpieces, chasing shooting stars
Favorite Food: Comet cotton candy
Favorite Color: Rainbow swirl

Buzzy Whirrington
Hobbies: Collecting space pebbles, doing robot dance-offs, telling cosmic jokes
Favorite Food: Moonberry pie with meteor sprinkles
Favorite Color: Galactic green

Zoomer Boltz
Hobbies: Racing through asteroid fields, stargazing with a telescope, teaching space yoga
Favorite Food: Rocket fuel smoothies
Favorite Color: Meteor orange

KNOW YOUR PLANETS

SPARKLEBOT'S PLANET
Name: Gigglesphere
Type: Whimsiworld
Moons: 3 shimmering dream moon
Length of Day: 20 hours of playful sunshine
Length of Year: 290 days of fanciful festivities
Populated by: Bounce-a-lot Bunnies and Tickleberry Pixies

BLOOPZILLIA'S PLANET

Name: Mirthville
Type: Quirktopia
Moons: None
Length of Day: 15 hours
of quirky surprises
Length of Year: 420 days
of merry madness
Populated by: Sillywiggle
Wigglesnouts and
Gigglesnort Gnomes

LUNA-DODDLE'S PLANET

Name: Whimsiwhirl
Type: Giggleverse
Moons: 1 laughter moon
Length of Day: 27 hours
of perpetual laughter
Length of Year: 365 days
of chuckles
Populated by: Giggly
Puffkins and Snickerdoodle
Sprites

GALAXI-GIGGLE'S PLANET
Name: Chuckleland
Type: Jollyverse
Moons: None
Length of Day: 19 hours of jollification
Length of Year: 280 days of merrymaking
Populated by: Chuckleberry Chortlekins and Guffawgrin Goblins

STARDUST SPARKLEPANTS'
PLANET

Name: Sparklepop
Type: Glitter Planet
Moons: Alot
Length of Day: 20 hours
(frequent sparkle breaks)
Length of Year: 400 days
(slow twirls around its star)
Populated by: Glitter
Gnomes and Twinkle Tots

AURORA BOREALIS'
PLANET

Name: Bounceberry
Type: Trampoline Planet
Moons: Five rubbery
bouncy moons
Length of Day: 12 hours
(constant ups and downs)
Length of Year: 300 days
(lots of leaps around
its star)
Populated by: Springy
Sprites and Jumping Jacks

QUIRKATRON'S PLANET
Name: QuirkStar
Type: Oddball Planet
Moons: Three mischievous moonlets
Length of Day: 25 hours (extra time for quirkiness)
Length of Year: 400 days (zigzagging orbits)
Populated by: Quirky Quokkas, Peculiar Pigeons, and Cosmic Cats

ASTRO-GIGGLES' PLANET
Name: Fizzytopia
Type: Bubble Planet
Moons: Four fizzy moon bubbles
Length of Day: 14 hours (bubbling with activity)
Length of Year: 270 days (bubbly orbits)
Populated by: Bubble Buddies, Sparkling Sprites, and Green Aliens

FLUFFYBOT'S PLANET
Name: Giggleopolis
Type: Silly City Planet
Moons: None (too busy with laughter)
Length of Day: 15 hours (filled with chuckles)
Length of Year: 350 days (countless comedy festivals)
Populated by: Giggling Goblins and Chuckle Champs

GIGGLE-PLEX'S PLANET

Name: Glimmerglobe
Type: Shimmer Planet
Moons: None
Length of Day: 22 hours
(shimmering moments)
Length of Year: 380 days
(sparkling orbits)
Populated by: Shimmering
Sprites, Gleaming Gnomes,
and Radiant Rabbits

ZIPPY-ZONK'S PLANET

Name: FunkyNova
Type: Groovy Planet
Moons: One giant disco ball moon
Length of Day: 30 hours (for non-stop dance parties)
Length of Year: 420 days (funky orbits)
Populated by: Groovy Goblins, Disco Dinosaurs, and Hip-hop Hares

GIGGLOPOLIS' PLANET

Name: ZanyZephyr
Type: Breezy Planet
Moons: Four twirling
tornado moons
Length of Day: 14 hours
(filled with gusty giggles)
Length of Year: 270 days
(winding orbits)
Populated by: Whimsical
Windmills, Playful
Pinwheels, and Zephyr
Zebras

THE END

GLOSSARY

Here's a glossary of some words used in this book that might be new to young readers:

ASTEROID: Rocks floating around in space.

CELESTIAL BEING: may refer to: A sky deity. An angel. Extraterrestrial life.

COMET: An icy rock that lets off gas and dust, which may form tails when it is flying close to a sun.

COSMIC: Relating to the universe and outer space.

GALAXY: A collection of thousands to billions of stars held together by gravity.

METEOR: The streak of light caused when a meteoroid enters a planet's atmosphere and starts to burn from the heat of friction.

METEORITE: A meteoroid that lands on the surface of a planet.

MOON: A natural object that travels around a bigger natural object.

NEBULA: A cloud of dust or gas found between stars.

ORBIT: The curved path that a planet, satellite, or spacecraft moves as it circles around another object.

PLANET: A large body in outer space that circles around the sun or another star.

SOLAR SYSTEM: A set that includes a star and all of the matter that orbits it, including planets and other objects.

STAR: A ball of shining gas, made mostly of hydrogen and helium, held together by its own gravity.

SUN: The star in the center of our solar system.

SUPERNOVA: The explosion of a star that makes it as bright as a whole galaxy.

UNIVERSE: All of space and time, and everything in it. It's everything ever!

This website is a wonderful resource if your child is curious about space and loves to explore. https://spaceplace.nasa.gov/

Dear Friends,

As we reach the conclusion of "Counting Cosmic Robots," let's take a moment to appreciate the importance of early learning and the delightful journey we've embarked upon. This book isn't just a story; it's a key that unlocks curiosity, cognitive growth, and a lifelong love for learning.

Thank you for being part of this interactive adventure. By counting, laughing, and exploring with your child, you've laid a strong foundation for their educational journey. Your dedication to making learning joyful and engaging is truly commendable.

As we bid farewell to our cosmic escapade, remember that the curiosity and enthusiasm for learning you've nurtured will continue to guide your child's path. This book is just the beginning of a lifelong adventure filled with exploration, growth, and endless possibilities.

Cheers,
Mister Jon
Panajachel, Guatemala 

We hope you've enjoyed the cosmic journey of "Counting Cosmic Robots" and that it brought joy and learning to your little one's world. If you found this book engaging and valuable, we would greatly appreciate it if you could take a moment to leave a review. Your feedback helps us continue to create educational and entertaining content for children.

And while you're at it, don't forget to explore our other delightful children's books, each crafted to ignite imagination, foster learning, and bring smiles to young faces. Thank you for being part of our literary adventure, and we look forward to sharing more stories with you in the future!

Thank you again!

Mister Jon
hola@mister-jon.com